POISON PROS!

Clever Critters

By Tracy Vonder Brink

TABLE OF CONTENTS

A Crabtree Seedlings Book

Crabtree Publishing

crabtreebooks.com

School-to-Home Support for Caregivers and Teachers

This book helps children grow by letting them practice reading. Here are a few guiding questions to help readers build comprehension skills. Possible answers appear here in red.

Before Reading:

- What do I know about this topic?
    - *I know that a sting from a wasp really hurts.*
    - *I know that some plants aren't safe for pets to eat.*
- What do I want to learn about this topic?
    - *I want to learn which animals use poison.*
    - *I want to learn why poison is helpful.*

During Reading:

- What have I learned so far?
    - *I have learned that using poison is an adaptation.*
    - *I have learned that adaptations help living things survive.*
- I wonder why...
    - *I wonder why poison frogs are the most brightly colored frogs in the world.*
    - *I wonder why a slow loris uses poison to keep other slow lorises away.*

After Reading:

- What did I learn about this topic?
    - *I have learned that a yellow jacket stings to protect its nest.*
    - *I have learned that a box jellyfish uses its tentacles to sting and kill fish.*
- Read the book again and look for the glossary words.
    - *I see the word **survive** on page 3 and the word **spines** on page 6. The other glossary words are found on pages 22 and 23.*

POISON PROS!

An **adaptation** is something that helps a living thing **survive**.

Making and using **poison** is one adaptation.

These animals and plants are poison pros!

A poison frog's skin is dangerous to touch.

Its bright colors warn **predators** to leave it alone.

Eating the frog will harm them.

Poison frogs are the most brightly colored frogs in the world.

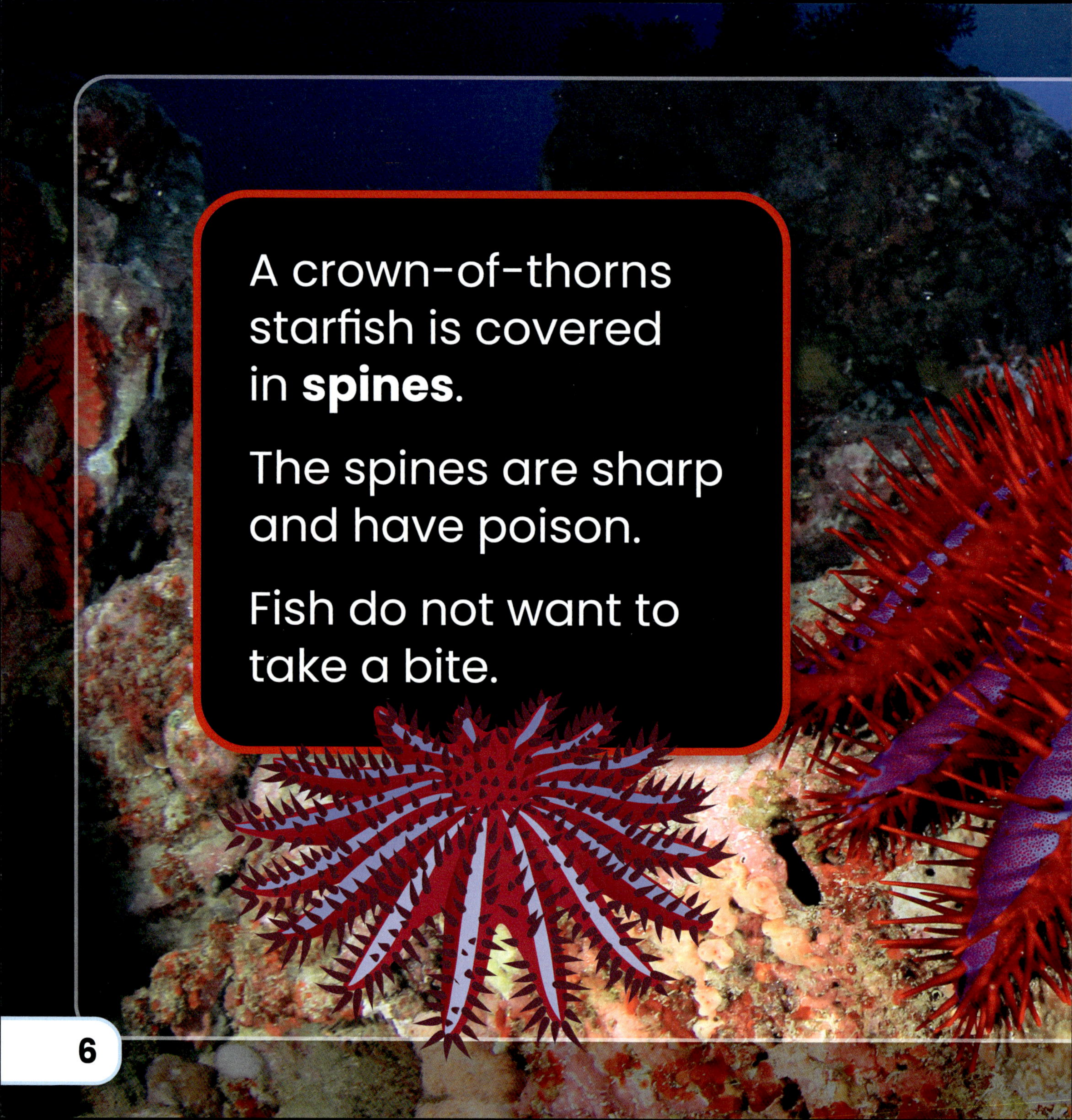

A crown-of-thorns starfish is covered in **spines**.

The spines are sharp and have poison.

Fish do not want to take a bite.

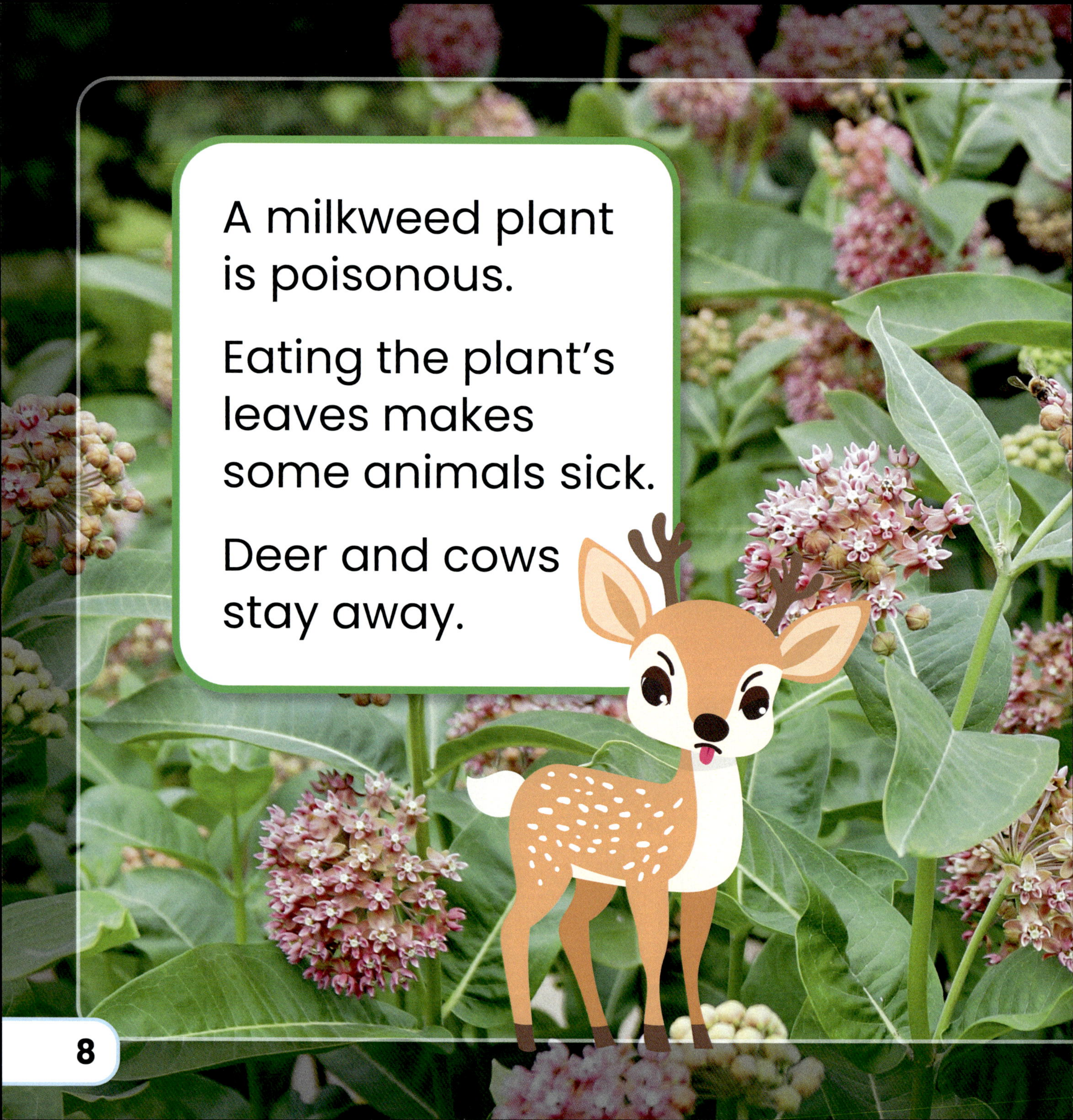

A milkweed plant is poisonous.

Eating the plant's leaves makes some animals sick.

Deer and cows stay away.

Not all animals stay away from milkweed. Some butterflies, bees, and other insects feed on milkweed flowers.

But monarch caterpillars have adapted to eat milkweed.

Eating milkweed makes the monarch caterpillar taste bad when it becomes a butterfly.

Birds do not eat the monarch butterfly.

A yellow jacket guards its nest.

It stings anything that comes near.

Its stinger carries poison that hurts.

The poison of stinging and biting creatures is called venom.

A slow loris has poison on the inside of its upper arms.

It licks the poison so that it gets on its teeth.

It then bites other slow lorises to keep them away.

A scorpion hunts a spider.

The scorpion has poison in its stinger at the end of its tail.

The scorpion stings the spider. The poison **stuns** the spider so that it cannot fight back.

A box jellyfish hunts small fish.

It grabs the fish with its **tentacles**.

The tentacles have poison that stings and kills the fish.

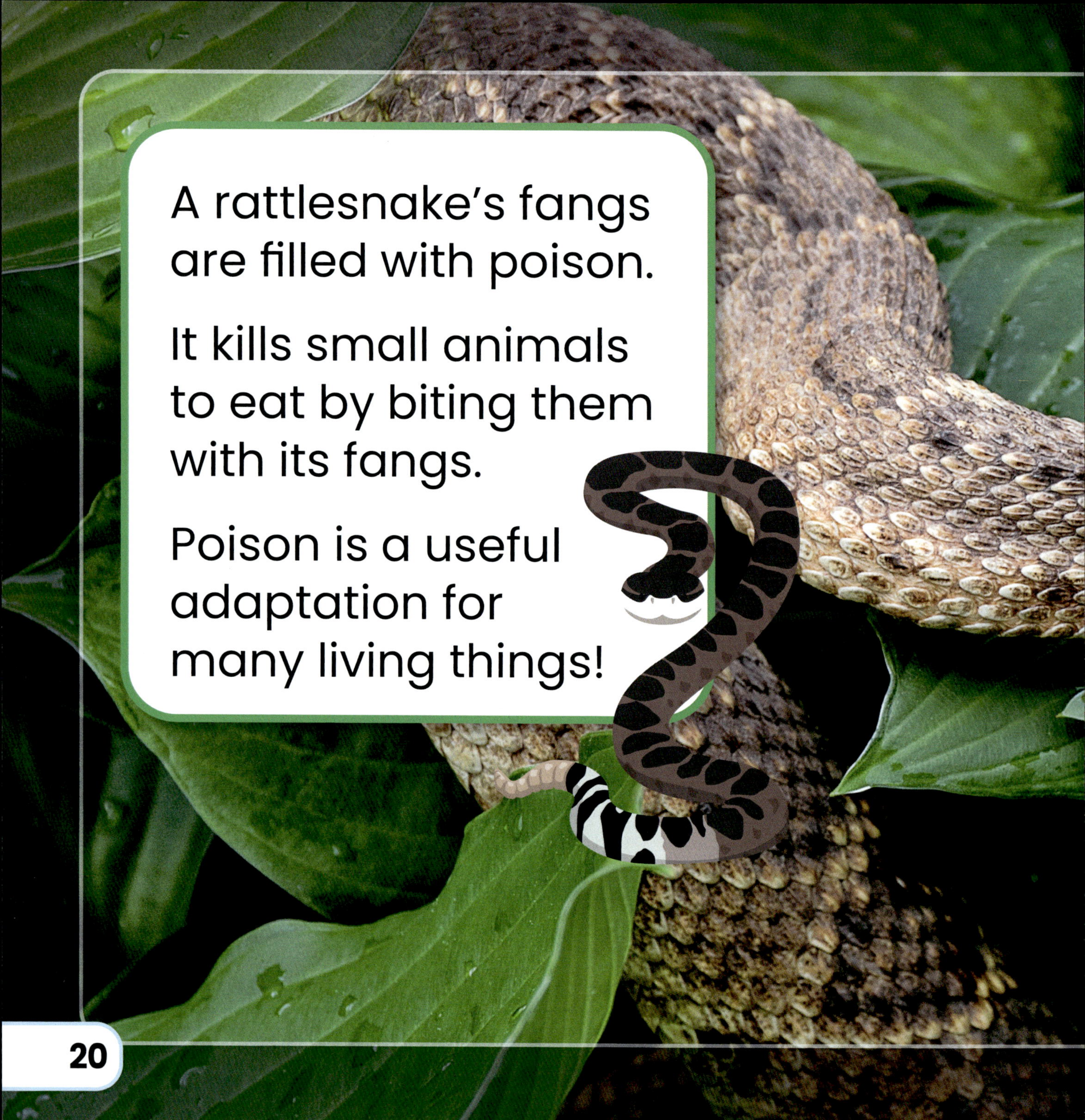

A rattlesnake's fangs are filled with poison.

It kills small animals to eat by biting them with its fangs.

Poison is a useful adaptation for many living things!

Glossary

adaptation (ad-uhp-TEY-shuhn): A change in an animal or a plant that helps it live in a certain place

poison (POI-zuhn): A substance that harms or kills an animal or person if it gets into the body

predator (PRED-uh-ter): An animal that lives by hunting and eating other animals

spines (spahynz): Sharp, pointed parts on an animal or plant

stuns (stuhnz): Makes something unable to react

survive (ser-VAHYV): To continue to live

tentacles (TEN-tuh-kuhlz): Long, arm-like body parts in some animals, used for grabbing things and moving

Index

About the Author

Tracy Vonder Brink loves to learn about science and nature. She has written more than 100 nonfiction books for kids. Tracy lives in Cincinnati, Ohio, with her husband, two daughters, and two rescue dogs.

Written by: Tracy Vonder Brink
Designed by: Margaret Salter
Proofreader: Melissa Boyce
Photographs: All images from Shutterstock

Crabtree Publishing

crabtreebooks.com 800-387-7650

Printed in the U.S.A./CP2025

Published in Canada
Crabtree Publishing
616 Welland Avenue
St. Catharines, Ontario
L2M 5V6

Published in the United States
Crabtree Publishing
347 Fifth Avenue
Suite 1402-145
New York, New York, 10016

Library and Archives Canada Cataloguing in Publication
Available at Library and Archives Canada

Library of Congress Cataloging-in-Publication Data
Available at the Library of Congress

Hardcover: 978-1-0398-8113-6
Paperback: 978-1-0398-8474-8
Ebook (pdf): 978-1-0398-8233-1
Epub: 978-1-0398-8353-6
Read-Along: 978-1-0398-8558-5
Audio: 978-1-0398-8516-5